G000026584

HARTLAND POINT TO NORTH FORELAND

Mike Smylie

AMBERLEY

Acknowledgements

Photographic books such as these are as much about the visual as the writing and therefore an author assumes the greatest gratitude to those who supplied the quality photos. Although some come from my own collection, for some of which I was behind the lens, many do not and therefore I am especially indebted to Jan Pentreath, John McWilliams, Nigel Legge, John Buchanan, Billy Stevenson, William Gale, Andy Maccloud, Robert Simper, Martin Ellis, Luke and Joanna Powell, Martin Castle, Jim Richards, Rudiger Bahr, Jonny Nance, Alan Toms, Bryan Roberts, Carol Williams, Richard Major, Peter Rees and the Cornish Maritime Trust, and lastly to the other individual members of the 40+ Fishing Boat Association, whose names I cannot remember, for their contributions over the years.

Secondly I must extend thanks to Rich Clapham of Falmouth for taking me oyster dredging in the *Holly Ann* and to skipper Jo Andrews of Newlyn for the ten days spent aboard the *Excellent* out in the Western Approaches several years ago.

General Note to Each Volume in This Series
Over the course of six volumes, this series will culminate in a complete picture of the fishing industry of Britain and Ireland and how it has changed over a period of 150 years or so, this timeframe being constrained by the early existence of photographic evidence. Although documented evidence of fishing around the coasts of these islands stretches well back into history, other than a brief overview, it is beyond the scope of these books. Furthermore, the coverage of much of today's high-tech fishing is kept to a minimum. Nevertheless, I do hope that each individual volume gives an overall picture of the fishing industry of that part of the coast.

For Ana and Otis

First published 2014

Amberley Publishing
The Hill, Stroud
Gloucestershire, GL5 4EP

www.amberley-books.com

Copyright © Mike Smylie, 2014

The right of Mike Smylie to be identified as the Author of this work has been asserted in accordance with the Copyrights, Designs and Patents Act 1988.

ISBN 978 1 4456 1455 7 (print)
ISBN 978 1 4456 1469 4 (ebook)

British Library Cataloguing in Publication Data.
A catalogue record for this book is available from the British Library.

Typeset in 9.5pt on 12pt Celeste.
Typesetting by Amberley Publishing.
Printed in the UK.

Introduction

In Volume 1 of this series I mentioned that only Cornwall could come anywhere near second place when considering fishing activity in Britain, first place being given to the fishers of the east coast of Scotland. Historically, Cornwall, whose coastline begins not far short of Hartland Point, had close access to the rich Western Approaches as well as its own coastal waters, which were full of fish of all types although it is only pilchards, mackerel and crabs that come foremost to mind today. Nevertheless Newlyn, at its western fringe, remains one of the two busiest fishing ports in England.

The other is of course Brixham in the adjacent county of Devon, a county that also has a history of fishing dating back to when ships sailed from ports such as Dartmouth, Plymouth, Teignmouth and Exmouth over to Newfoundland (sometimes referred to as Devon's furthest west) after John Cabot's voyage of 1497. Brixham has the accolade of being the first port to develop trawling in the North Sea, although the Essex creek of Barking sometimes claims the title. However, to step back a few years, it is deemed to be the men of the small beach and village of Beer that taught Brixham how to trawl, and thus claim the title. Of course, this isn't always good publicity as there are many who point out that trawling was responsible ultimately for the disastrous state of the fishing industry in Britain and that the blame lies at the feet of these folk. Given that an early 'trawle' called a wondyrychoun was in use in the Thames in the fourteenth century, perhaps it would be best if the blame was laid elsewhere!

At the other end of the English Channel, sheltering under the North Foreland, lies Ramsgate, once a prolific sender of trawling boats into the North Sea. In between the two extremes are numerous fishing stations. Perhaps the best known for its surviving vibrant beach-based fishery is Hastings, where activity on the Stade continues against the ever-invading march of European legislation. But towns such as Eastbourne, Brighton, Bognor and Bournemouth all emerged from small fishing hamlets after the start of the rush for sea-bathing in the seventeenth century. And between these are lots of small hamlets that have remained peacefully small though just as active by way of the fish landed for local use.

The range of fish is vast: Richard Carew, in his *Survey of Cornwall* of 1602, for example, tells us that salmon, trout, pilchard, mackerel, flatfish, rays, thornbacks, soles, flukes, dabs, plaice, sprat, smelts, whiting, scad, chad, sharks, cuttlefish, eels, bass, mullet and hake were all common. On top of that were winkles, limpets, cockles, mussels, shrimps, crabs, lobsters and oysters. Add scallops and remove some of the unpalatable types and you've a picture of today's fishing activity.

In terms of landings into England, Brixham tends to lead over Newlyn, though the latter lands more bottom-feeding (demersal) fish. The figures for 2012 are that Brixham landed 4,800 tons of demersal fish, 3,000 tons of pelagic fish and 7,800 tons of shellfish (total 15,600 tons), while for Newlyn figures for the same groups are 6,200 tons, 2,100 tons and 2,500 tons, giving a total of 10,700 tons. However, perhaps the most surprising fact about the 2012 figures is that Plymouth landed 15,700 tons in total, overtaking even Brixham. Most of this seems to have been pelagic fish – mackerel in the main – much of which is processed. Mind you, compare this to Peterhead's landing of 105,700 and you begin to see the difference! In terms of value of the catch, Brixham was ahead at £27 million, with Newlyn at £20.3 million and Plymouth at £16.1 million. Fourth place in the annual statistics goes to Shoreham, mainly due to its scallop landings, which exceeded those of Brixham by 100 tons. However, take into account Plymouth's scallop landings and the total for the three ports is 10,758 tons of scallops, which is well over half of the tonnage landed into England and exceeds that of the Isle of Man scallop industry (contrary to what was stated in error in the introduction to Volume 3).

The south coast has a diverse collection of fishing boats, although there are similarities in that most traditional craft were transom-sterned. Generally, the lug rig was favoured by the inshore boats whereas the larger trawlers of Devon adopted the gaff rig. To the east the rounded Sussex workboats reflect the craft seen across on the French coast, while the western luggers themselves have similarities to those of the Breton coast. Vessels evolved through usage and with that usage being common to both sides of the Channel, it's not surprising that French and English boats have this common thread in their pedigree.

Finally, as I write this, as the west and south coasts are being bombarded by a constant stream of winter storms that are causing havoc with coastlines and river levels alike. The fishing fleets, unable to sail, are tied up with their crews receiving no earnings. The news might be full of promises of no expenses to be spared and 'money will be no object' for flood victims and farmers, but do we see any chance of fishermen being reimbursed for the great hole in their pocket? Hammered through oppressive EU legislation, suffering from the highest death rate among workers, and working in some of the most dangerous and cold conditions imaginable, fishermen are generally overlooked except in the knick-knack shops of tourist villages. Yet their contribution to the growth of the country over several centuries has been massive – their place in the Royal Navy and the growth of the Empire, their contribution in both world wars – as is of course their feeding the country with endless supplies of fish. Theirs is a heritage to be proud of, and a tiny bit of it, I hope, is portrayed in this series of books.

Fishing Ways

The English Channel as a whole, which includes some 90 per cent of the coastline represented in this volume, is often overlooked as a serious commercial fishing ground. Nothing could be so wrong for it is teeming with fish. Part of the problem with identity might be the fact that, in the days of territorial waters, much of the fishing ground was shared between Britain and our French neighbours. But a strong fishery there has been and over the years fishing has occupied a great number of folk. In the main these include a strong tradition of pilchard and mackerel fishing in Cornwall, as seen in the introduction. Pilchard seining was a traditional artisan fishery, involving seine companies employing many men, while drift-netting for pilchards overtook seining as the preferred method for Cornish fishers after about 1870 until the stocks declined. Mackerel were caught in drift-nets in vast numbers, but hand jigging today is a more environmentally friendly way of landing fresh mackerel, much of which is smoked in West Country smokehouses.

There has been, and still is today, a major lobster and crab fishery. Cornwall is also home to the only remaining fishery worked under sail – the Falmouth oyster fishery. For over 500 years the waters of the Fal and Truro rivers have been fished by sailing boats or under oar, and this continues in the same manner due to a local by-law of 1876 and certain EU restrictions. In the shallow water small punts are used to drag a dredge across the beds, but in the deeper water larger boats are used and these are allowed to have an engine to travel between their home port and the dredging grounds. Most boats used are what are termed Falmouth Working Boats and it is believed that they are members of the world's last oyster dredging fleet working under sail alone. One of the interesting facts about the Falmouth Working Boats is that it is only the name and the work that ties them together as they are an assemblage of some very different craft. Having said that, many of the locally built craft are similar in that they are transom-sterned and up to about 30 feet in length, but probably a bit shorter. These have long, deep keels, full bilges and deep forefeet, with the maximum beam amidships and many have a small cuddy for shelter. Other boats have been brought into the area to work the dredges, some successfully and others not, due to their shape. Nevertheless, the fleet has been known as one of the most diverse in Britain in the sailing era. One boat, the *Zigeuner*, FH89, was famously built in Germany, rebuilt in Restronguet in the 1840s and fished up to the 1940s.

The fishery still survives – just – though pollution, the stress of the job, retirement among the older fishers and unwillingness among newcomers to persevere has seriously depleted the number of boats working the dredges to a mere handful.

Fibreglass versions of these craft have been built and many of these, along with wooden boats, with a heightened rig, race out of St Mawes during the summer months when there's no oyster dredging. When working, Falmouth Working Boats carry about 300 sq. ft of sail, but when they race this increases to the voluntary limit of a massive 1,000 sq. ft. Most have no engines and are very handy boats to race.

Chesil Beach was once renowned for its mackerel seines while Selsey Bill and Bognor Regis were once home to a healthy lobster and crab fishery. Today, although nothing like the fisheries of former times, both continue to be fished. To the east, Hastings has managed to hold onto its sole fishery, which was accredited by the Marine Stewardship Council as a sustainable fishery (and given its blue logo), on top of which the local herring and mackerel fishery achieved the same status.

Gill netting for hake was, until a few years ago, practised by some fifty boats working out of Newlyn. Many of the first set of photographs were taken by the author on one of these trips back in the 1990s.

Gill-Netting for Hake

The *Excellent* was built in 1931 as the *Efficient* in Sandhaven, near Fraserburgh, as a typical large herring drifter, and fished locally until being sold to Newlyn just before the Second World War. She was requisitioned during the war and returned to fishing soon after. Here she is in 1952, rigged for trawling. (*Billy Stevenson*)

At the time of the author's ten days aboard her, she was the oldest first class fishing boat in the British fleet. In this photo she is rigged up with sails for a photo opportunity though the winch for gill-netting can clearly be seen, as can the net pounds at the stern. (*Billy Stevenson*)

The net is being shot under the port stern quarter, along a channel leading from the net pound forward of the wheelhouse.

Although the net here is being redd into the pound, the photo illustrates where the pound is and the channel through which the net travels when being shot is clearly seen on the left.

The hauling process has started with the net being redd back into the pound.

Hake get trapped by their gills in the net, hence its name.

Hake in the winch gets a shout of 'hake-o' before it gets mangled!

A good example of a hake coming over the top of the winch. Unlike trawling, the crew aren't faced with a load of fish coming aboard at one time and the rest of the crew are ensuring the net gets redd into the pounds correctly while one tends the winch.

Sometimes extracting the hake from the net is simply a matter of a twist and it's free.

At other times it can be a long process of extraction.

The net is aboard and has been shot in another position while clearing the deck takes place. Mackerel (seen on deck under the winch) are a problem because they need to be removed from the net though they are not worth keeping as they will have gone off by the time the vessel gets back to port in a week.

Gutting takes place as soon as possible to ensure freshness of the fish when iced up.

Fishing, and gutting, takes place night and day. The crew grab sleep just as best as they can when the decks are clear and before the net is hauled once again.

Boxes of fish ready for lowering in to the fish hold below.

Each fish is carefully stowed in fish boxes with plenty of ice in the fish hold.

After a hard day's work, ample supplies of food are served up in the tiny mess-cum-galley behind the wheelhouse, served in dog bowls! One thing about this boat was the food which was excellent – roast meat every day, usually cooked by skipper Jo Andrews on the left.

The sleeping quarters in the aft cabin. Mine was the one in the centre of the photo, right above the propeller, ensuring a disturbed sleep. Getting in and out wasn't easy either!

Pilchard Seining

A view of a large seine of pilchards at Porthgwarra, near St Levan. The net is laden with pilchards, has been towed to shallow water, and is being emptied using wicker baskets called dippers in the seine boat that was pulled into the seine. Many thousands of fish would be caught in one seine and the fish could remain fresh in the net for several days while anchored to the seabed. (*Bryan Roberts*)

Seine-net boats at Sennen Cove loaded with the haul. Note the shape of the seine boats and just how many are lying on the beach. (*Bryan Roberts*)

The remains of one of the last seine boats at Sennen Cove in 2005.

Pilchards aboard the *Prevail*, having been caught in a ring-net *c.* 2001. (*Martin Ellis*)

Lobsters and Crabs

Withy ink-well pots in Cornwall: this is the traditional type of pot, which is seldom used today, the square metal pot being preferred.

First day issue of stamps showing both withy and steel framed pots. This is dated September 1981 and celebrates the centenary of the Fishermen's Mission.

Traditional withy pot makers do still exist. Here, Nigel Legge of Cagdwith is in his workshop making one of his small withy pots that he sells to passing holidaymakers. (*Nigel Legge*)

On the other hand, Nigel also makes full-sized withy ink-well pots, some of which he sells and some with which he fishes. Most of the withies come from Somerset though some are grown in Cornwall. (*Nigel Legge*)

Most of the pots he uses these days, however, are of the metal half-round shape. Here he is hauling aboard his small open boat which he works off the beach at Cadgwith where his father worked before him, and his before him. (*Nigel Legge*)

The pots have been re-baited and are now being put over the side. Nigel was the 'guide and mentor' to Monty Halls, the fisherman's apprentice in the BBC2 series of the same name filmed in and around Cadgwith, and broadcast in 2012. (*Nigel Legge*)

Withy pots on the seafront at Brighton. 'SM' represents the registration letters for Shoreham-on-Sea.

Oyster Fishing Under Sail

The Falmouth Working Boat *George Glasson* drifting sideways down the tide over the Carrick Roads oyster beds in 1974. She was built in 1898 in Porthleven as a lug-rigged pilchard driver and was half-decked. She started oyster dredging in 1923, at which time she was converted to a gaff cutter, a rig much more suited to dredging. The boat is reefed and the foresails lowered while the men work the dredges on the port side. (*Robert Simper*)

Here, one of the Falmouth Working Boats is dredging in 2014 in very little wind. The sail is set so that the boat sails at a dead slow speed while the tide takes the boat along with it, pulling the two dredges over the banks. The engine is used only to travel between mooring and fishing grounds and cannot be used to sail upstream or upwind. The boats are renowned for having high-peaked gaffs.

Above: Most boats carry two licensed dredges, the current cost being £165 a year for each one. Here, aboard the *Holly Ann*, Richard Clapham is shooting one dredge, the other being down, so that they are hauled in succession.

The dredge being hauled. Each one spends some ten minutes in the water – that being the time taken to haul in, empty, re-shoot and sort through the catch of each dredge.

Richard is hauling in. If the dredge is full, there's a fair weight in the dredge and a good deal of effort is needed to lift the mouth of the dredge onto the wooden platform.

There's a water bailiff who keeps an eye on the fishing. Undersized oysters are thrown back in, their size being dictated by a brass ring which is 67 mm in diameter – any oyster should be able to sit on the ring. However, those that are border-line – i.e. they will just sit but are thin – are often returned to fatten up. Fishing is maintained by the attitude of the fishermen, the wind for sailing and the weather in general.

Here, the dredge is being emptied into the wooden tray. The aft dredge is still down and the thin twine acts as a preventer in the event the dredge snags. The twine breaks and the buoy attached to the end of the dredge line drops overboard so that the whole dredge can later be reclaimed.

The by-catch consists of scallops, queenies, mussels, starfish, tiny crabs, weed and slipper limpets. Some of the fishermen, including Richard, collect the invasive slipper limpets to take ashore to protect the oysters from attack. The starfish also have a nasty habit of forcing shellfish apart a tiny bit, extending their stomachs into the gap, and eating the contents, so they are often taken ashore for disposal as well. They make good garden compost I'm told!

Outside of the River Fal, offshore boats dredge for oysters and scallops. Here, on the quayside at Mylor, are steel beams with dredges attached. These run over the seabed on the wheels and the dredges gather up everything in their sight. These dredges can do a terrific amount of damage to the ecology of the seabed and are only allowed in certain areas.

Seine-Netting in the Solent

Small boats such as this worked around Southampton Water and out into the Solent, using seine, trammel and peter nets in shallow water. Here, a seine is being worked off the shore.

Above: Basketing fish from a net into the boat to transport it back to Portsmouth market. Herring and sprats were common in autumn, mackerel in summer. At other times the boats long-lined, netted mullet and fished for whelks.

Two Portsmouth seine-net boats – *Harriet Kate*, P63, and *Snowdrop*, P65 – photographed at the International Festival of the Sea at Portsmouth in 2005.

Fishing Boats

Again, as in the previous volume, we see a diverse collection of fishing craft working these shores – from Cornish luggers to Ramsgate smacks. In between, we find many beach-based craft such as the various Cornish crabbers, the Devon luggers, the Chesil Beach lerrets, the Selsey crabbers and the Sussex and Kent beach boats. All have their roots, which suggest a tinge of French influence. In Cornwall, many of the traditional types were transom-sterned, although around the western tip we see double-enders. These double-enders have many similarities to the luggers of the East Coast of Scotland and, although the Cornish and Scots mingled in various fisheries, the suggestion is that both types evolved unconnectedly simply through their mode of work i.e. through the necessities of drift-netting, the Scots for herring and the Cornish for mackerel.

Boats did travel back in the days when influences were playing upon evolution: the first Cornish boats to Ireland in 1816 and to the East Coast of Scotland via the Forth & Clyde Canal possibly in 1826, although this date might be a bit early. East Anglian boats came to Cornwall after 1860 when the railway had been completed from Cornwall across Brunel's Tamar Bridge, which opened the year before, to the rest of England.

We also know that most fishing boats were mostly undecked until the middle of the nineteenth century. One exception was the sprit-rigged Brighton hog boat or hoggie, which was a bluff-bowed, rounded-hulled clumsy-looking affair, ranging up to 35 ft in length. The drawings of Edward Cooke are largely the only visual source of these craft remaining as they were well into decline by the 1860s, with the luggers taking preference.

Brixham trawlers travelled to the North Sea to open up the fishing there and we shall learn much more about that in the next volume. It suffices to say that they were fishing the North Sea before the beginning of the nineteenth century and that it was a Brixham boat that discovered the famous Silver Pits in the late 1830s. However, many Devon men will tell you that it was Beer men that made Brixham; these fishers in small luggers working off their exposed beach deciding on larger boats which they then based in Brixham, across the bay where there was shelter and a deepwater mooring.

Kent men were so impressed with the Cornish luggers that many were specifically built by the renowned Cornish builders and sailed eastwards. One fine example is the 1905-built *Happy Return*, which was ordered by a Folkestone fisherman to replace an earlier vessel. She worked there until 1968, when she moved to Kings Lynn and was renamed *Britannia*. She was decommissioned in 1998 and was threatened with being scrapped until she passed to the Mount's Bay Lugger Association in Penzance, who rebuilt her and returned her to her original name.

Double-ended luggers at St Ives in about 1900. It's high tide and some are moored alongside Smeaton's Pier. Strangely, the pilchard boat alongside the quay has withy crab pots aboard: usually these were fabricated from metal in St Ives.

Mousehole harbour in about 1880 with dried out double-ended luggers and some smaller crab boats. The large boat, *Fearless*, 150PZ, was a mackerel lugger while the smaller decked one in front drifted for pilchards.

The Fowey-registered *Silver Wave*, 158FY, dried out at Fowey. Generally East Cornish boats were transom-sterned against the double-enders to the west of the county. (*Dellon Boothby*)

Another transom-sterned Cornish vessel, *Two Sisters* of Fowey, in calm conditions. She is undecked and typical of the smaller gaff-rigged boats working out of Polperro, where they were called Polperro gaffers. Similarly shaped craft from Plymouth were referred to as Plymouth hookers. However, most fishermen simply called them 'boats'!

The Polperro gaffer *Viloma May*, FY20, sailing goose-winged at the Looe Luggers Regatta in 2009. Built in 1896, since ceasing fishing she has sailed across the Atlantic under the ownership of Chris Rees and his partner Martha, up to Boston, and as far north as Greenland before returning to Cornwall.

Porthleven in the late 1940s. Although the boats all appear to be of the East Cornish type, the exception is the one double-ender PZ1 which is in fact the Lochfyne skiff *Perseverance*, once owned by the author.

A view of St Ives, with the deck view of a typical lugger. The boat is decked over, with a large fish hold covered with boards. The wheel and drum is an 'ironman', an early capstan, which made the job of hauling in the nets much easier and led to more nets being set.

For comparison, the deck view of the restored St Ives lugger *Barnabas*, SS634, built by Henry Trevorrow above Porthgwidden Beach, St Ives, in 1881. Built as a small mackerel lugger, she fished until 1954, after which she was converted to pleasure. She now belongs to the Cornish Maritime Trust and has been completely rebuilt with finance from the Heritage Lottery Fund.

The harbour at Folkestone in the 1950s with the lugger *Happy Return*, FE 5. The wheelhouse was added in the early 1950s and the rig has been removed, although the masts were retained as hoists. (*William Gale*)

Luggers at Looe; the lugger *Our Boys*, FY221, is coming alongside the east side of the river, where the fish market is. The small pill-box wheelhouses were a trend throughout Britain after the introduction of motors after about 1910, their advantage to fishermen being obvious. (*John McWilliams*)

Another lugger, this time *Snowdrop*, FY104, photographed in August 1965 at Mevagissey, with pilchard nets aboard. The mizzen sail has been retained to steady the boat while lying to the nets. (*John McWilliams*)

Snowdrop sailing across Douarnenez Bay in 2008, rigged once again in the traditional way, after attending the Brest Maritime Festival. Built in 1925 at Porthleven, she worked until the 1970s, then becoming a pleasure boat.

The Port Isaac lugger *Mizpah*, with its owner Capt. George Bennett standing alongside in about 1895. This boat was typical of the small beach boats that worked from the three 'ports' of Port Isaac, Port Gaverne and Portquin up to the Second World War. They worked crab pots in summer as well as hand- and long-lining. In autumn and winter they drift-netted for pilchards and herring. (*Martin Castle*)

Herring boats at St Ives. These boats, more commonly called the pilchard drivers, fished both pilchards and herring. Here one man is using a sweep – the long oar – although the sails are set. The mainsail appears to be a shortened sail.

Herring gigs at St Ives. These were a smaller version of the pilchard drivers, heavily built. They were able to get out to the fishing grounds quicker than the larger boats though many of the older boats seemed to have been wrecked, which made them unpopular. Nevertheless they survived the advent of motorisation and some continued fishing as motor gigs, though by the 1940s they were hardly recognisable, with capstans for crabbing and trawling. Latterly most were painted light blue. Although these gigs worked from various north Cornish ports, they are not to be confused with the pilot gigs.

The St Ives jumbo *Celeste*, built by Jonny Nance, the first of the small replica boats to sail out of St Ives in many years. Again, these boats were smaller versions of the larger pilchard drivers and were built in place of the gigs, which the fishermen weren't keen on. The name came from a public outcry after the sale of an African elephant, a popular attraction in London Zoo, to the US in 1882. (*Jonny Nance*)

Mevagissey toshers. These small boats – 19 ft 11 in. in length – were built as a result of avoiding the hike in harbour dues for vessels over 20 ft in length. These were rigged with a single mast, either lug-rigged or as pole-masted cutters. Both these boats in the foreground have a tiny foredeck with storage below while a few had a cuddy for the crew to shelter in.

Pioneer, PZ277, sailing in St Ives Bay in 2007. Built in 1899 by William Paynter in St Ives, this boat was fitted with a steam engine and worked as a drifter. She was lengthened at one time from 35 ft to 47 ft and worked all manner of fishing, latterly trawling. The steam engine was replaced with a Kelvin in 1947. By 1991 the boat was derelict until bought and restored by Jim Richards.

Three luggers at Mousehole in 2010. On the left is *Happy Return* and alongside is the counter-sterned *Ocean Pride*, PZ134, built in 1919 by Henry Peake of Newlyn as the last boat built on Tolcarne Beach. She worked out of Mousehole, one of only two counter-sterned boats there. On the right of the photo is *Pioneer*, PZ277.

Luggers on the beach at St Ives. On the left is *Ebenezer*, SS340, a lugger built in 1867 and still fishing in 1930, but which sadly ended her days rotting away at Lelant. In the middle is *Gratitude*, a boat that eventually had three engines, another shaft emerging at an angle just forward of the quarter shaft seen here. On the right is a transom-sterned boat from around Land's End.

A typical Sennen Cove crab boat which probably drifted for pilchards at times. The bumpkin is nearly as long as the boat!

Minerva, FH58, a Cadgwith crabber built in the village in 1935 with an engine. The full rig was retained as engine reliability still was not good in deepest Cornwall at this time. (*John Buchanan*)

Antionette, FH626, a motorised Cornish crabber from Porthoustock, photographed about 1998. The fuller form than the previous two is obvious as this boat was only built for engine power. Porthoustock is a fishing cove that became dominated by jetties used to load road stone onto ships from nearby quarries. By the 1960s the beach was once again occupied solely by fishermen.

The Hope Cove crabber *Progress*, built in 1926 by the Jarvis brothers, which fished from the small harbour.

Across the Channel, the boats of the Channel Islands display influence from France. Here the small Jersey crab boat *Fiona*, 108J, resembles many of the Breton small boats though few of these are gaff-rigged, an influence that came from England.

Below: The schooner-rigged Guernsey mackerel boat here appears to be a throwback from the days of three-masted luggers such as the French bisquines. The third mast has been dropped and the gaff rig adopted. The hull shape is both French and southern English. (*F. W. Guerin*)

The boatbuilding company of C. Toms of Polruan was still building wooden fishing boats when this boat, *Eilidh*, BRD149, was launched in 2001 for owners in the Outer Hebrides. (*Alan Toms*)

Scottish boats also found their way to Cornwall. Here the traditional Scottish canoe-sterned St Monans-built *Gratitude*, Ba148, launched in 1950, was working from Looe under local ownership in 1999.

Carradale ring-net boats in Cornwall when fishing for mackerel in 1975. (*John McWilliams*)

A view of the harbour at Folkestone with a couple of Cornish-built luggers and a Scottish-built boat, *Fair Chance*, FE20, built in 1955 at Gardenstown and still fishing in 1997 though ceasing sometime soon after.

Above: Cornish luggers *Barnabas* and *Happy Return* off Looe in 2009, depicting the difference in size of these two boats.

A three-masted lugger sitting on the beach at Beer in 1877. This elm-on-oak-frames clinker-built lugger was built locally and was typical of the time. These were the forerunners of the later Beer luggers. Note the forward bumpkin to set the lugsail on.

Beer luggers *Nuddy* and *Butterfly* sailing off Beer. There used to be some twenty-five luggers up to 30 ft in length working off the beach but now the luggers are used to racing, though numbers have swelled in the last decade or so. Note the forward bumpkin again.

Beer may have made Brixham but Brixham developed these powerful trawlers which this photograph alludes to. Later the boats adopted the ketch rig.

The Brixham trawler *Pilgrim*, BM45, which has recently been rebuilt with Heritage Lottery funding by Ashley Butler at his Pill Creek Boatyard. Other boats to survive are *Leader*, *Torbay Lass*, *Vigilance*, *Provident*, *Deodar*, *Regard* and *Ethel von Brixham* while *Little Mint* is the last surviving cutter-rigged trawler. *Pilgrim* is the oldest surviving of these boats, built by J. W. & A. Uphams of Brixham in 1895. (*Bill Wakeham*)

Two trawlers, *Keewaydin*, LT1192, and *Vigilance*, BM76, sailing in Tor Bay during the annual Brixham trawler race. Although registered in Lowestoft, *Keewaydin* was built in Rye while *Vigilance* is the youngest of the surviving trawlers, built by Uphams in 1926.

Beach boats on the Devon coast, registered at Exmouth. Boats like these from Beer worked anywhere between Teignmouth and Lyme Regis. Again, the bumpkin, to which the lugsail is set, can be seen.

This little boat was spotted in 1998 at Axmouth. It has adopted the lute stern, resembling Sussex craft, and has a winch to haul pots.

A Chesil Beach lerret. Supposedly originating in the fifteenth century, the lerret evolved as a double-ender because of its ability to be readily launched and beached on the pebbly Chesil Beach. The Fleet is the stretch of water that lies inshore of the beach between Abbotsbury and Portland, and the boats were often rowed across this then carried over the bank of the beach into the sea. The Fleet trow later evolved for this task. The lerret had to cope with the surge of the waves onto the beach and the cross-current, and the double-ender, pretty much the same shape either end, was found best to deal with this. They were also perfect for smuggling, many sailing over to the Channel Islands to 'free-trade'.

By the early part of the twentieth century there were some forty boats moored up off Poole town, when they had been moved by the local council, who wanted their previous home at the quayside at Hamworthy to be redeveloped. Their size increased up to 25 ft, their floors became flatter and the forefoot was rounded to produce a handy form of boat. As they sailed even farther afield, sizes reached up to 30 ft, and by the 1930s they were still some thirty boats fishing full-time, mostly having adopted Bermudan rig. By 1936 there were only six boats left, and by the following year, only one, the *Polly*, PE69, remained. (*E. Bristowe*)

During the eighteenth century, Itchen Ferry was a small fishing village on the banks of the River Itchen, opposite Southampton. Small sprit-rigged craft worked off the beach, fishing the waters of the Solent and Southampton Water. By 1900 the boats were up to 30 ft, and had long straight keels with a slight rake and had adopted the cutter rig. The boats were three-quarter decked with a small forecastle under the foredeck that stretched one-third of its length. In the cuddy were two berths, a cupboard and a coal stove. The cockpit, or stern-sheets, had narrow side decks, and in front was the fish-hold. The boats spent most of their time fishing for shrimps and oysters before racing back to land their catches. This one, *Wonder*, SU120, was built in 1860 but has since been rebuilt.

Here the long keel and deep hull of the Itchen Ferry can be seen. The boat is *Freda*, CS110, built by Fay of Southampton in 1890. She was built as a mirror-image of *Wonder* using lightweight materials for Captain Sam Randall, a renowned yacht racing skipper of that era. Although she worked around the Solent most of her life, she also was a race winner. (*Earle Bloomfield*)

Referred to as a 'Selsey fishing boat' and registered at Littlehampton, this was more likely owned in Bognor (hence the name Bognor lobster boat) though it was probably based in Littlehampton, the only natural harbour between Selsey Bill and Shoreham. These long-boomers were cutter rigged and one such boat, *Gwendolene*, Li49, was built by old Granby Hopkins in 1893. These boats also dredged oysters and fished for mackerel while some were said to have a lucrative business free-trading brandy across from France!

The Brighton 'Hoggie' or 'Hog-boat' emerged as the common boat, there being some seventy such craft around the close of the century. Rigged with spritsails, the hoggies were unique to the town, although some were said to have been based farther along the coast at Shoreham. These hoggies were fully decked, unlike many other fishing boats of that era and were up to about 14 tons around 1789, and up to 35 ft (10.7 m) overall. They had a forecastle for living quarters, although none seem to have had a stove, as many of boats of that era with accommodation did. (*Edward Cooke*)

Coming onto the beach at Hastings. The boat is *Industry*, RX94, built in the town in 1870 and worked until the mid-1940s. Despite attempts to save her, she was burnt on Guy Fawkes Night in 1949, a fate that awaited many of Britain's wooden traditional boats.

The beach at Hastings. The lute stern belongs to the lugger *Happy Return*, RX198, which was built in the town in 1903 and sunk by a mine in Rye Bay in 1940. The lute stern was intended to break the wave as the boat came onto the shore.

Launching the *Mary*, 52RX, at Hastings by pushing her into the sea while the rudder is raised. Manhandling several tons of boat was not easy! (*George Woods Collection*)

A typical Hastings lugger on the beach displaying her rig. These boats were called '28 boats' or 'bogs' as they fell into the 28–30 ft length category. These boats trawled, drifted for herring and even went to the Yarmouth autumnal herring fishery.

A modern-day Hastings boat coming onto the beach in 2001. Here, the crewman prepares to jump ashore.

The cable from the winch or tractor is attached to the forefoot of the boat and the hauling process can begin. Note the runners on the hull to keep the boat upright although the hull is much fuller than the sailing boats.

Two boats on the beach: the elliptical-sterned *Our Pam and Peter*, RX58, and lute-sterned *Bloodaxe*, RX37. The elliptical stern was introduced in 1892 with the building of *Clupidae*, RX, following several successful designs on yachts. The theory was that the elliptical stern lifted with the wave. (*Rudiger Bahr*)

Above: Hastings beach remains a working beach, despite EU legislation that makes the job of fishing more and more difficult. Instead of attacking the jobs of these coastal fishers, legislation should be directed at those who do most damage. Small-scale fishermen account for some 95 per cent of fishers worldwide yet in Britain they only have 4 per cent of the fishing quota. Surely it is time this was changed?

A Ramsgate-registered boat, R300, which appears to be a Thames bawley, which seems an appropriate time to complete the fishing boats of this coast, which will be then continued in the next volume of this series.

Fisher Folk

Whereas fishing methods and fishing boats are the tools of the fishermen, it is the social history of the fisherfolk themselves that in some ways is the true story of fishing. These fishers survived in their insular communities strung out along the coasts of southern England and each had their own traditions and ways of living and fishing. Not until twentieth century fishing displaced these communities in favour of large ports did they change over generations. Son followed father who had followed grandfather into fishing – there was generally no other choice.

Throughout much of Britain fishing communities tended to be separate from the general populace, often set aside at the end of the town. Often these communities weren't just about fishing, and boatbuilding was an equally important part of the story. The boats had to be built and the south coast was home to several itinerant boatbuilders who would travel to a particular settlement to build a boat for some individual. Boatbuilding yards – or more likely available waterside fields – spread in the nineteenth century so that a whole host of builders were producing small boats for the local fishers. Sailmakers and blacksmiths contributed, as did the various net-makers. Once the fish were landed, they had to be either sent straight to market or hawked about the locality or processed. Processing usually involved either a salt-cure or a smoke-cure until the advent of refrigeration, deep-freezing and processes such as fish-finger production.

Brighton and Hastings are two prime examples of how fishermen were treated by those in control of the burgeoning towns. In Brighton they were moved to the very edge of the town as it was growing in both size and stance as sea bathing gained popularity. In Hastings, after decades of trying to oust the fishermen from the Stade, a Reconstruction Committee looked into the direction of the town in peacetime. They employed Mr Trystam Edwards to produce proposals, part of which involved the building of a new fishermen's quarter by the Stade, contrary to previous plans. When he put his proposals forward at a council meeting they promptly sacked him and employed Sidney Little, the incumbent Borough Engineer and Surveyor, who had already transformed the face of Hastings before the war and who was sympathetic to the previous order of minimizing the presence of the fishing community. Thus the Council's appointment of Little almost ensured the continued post-war destruction of the Old Town and its way of life. Much of the area around Rock-a-Nore was due to be demolished, the net sheds moved and restrictions laid on the fishermen. Thankfully the Council withdrew its hastily put forward plan after complaints from angry fishermen, townspeople and councillors who had only three days' notice of the plan. But this was only one example of the various battles, some legal, some not, the fishermen had with the Council over the years. Steve Peak's *Fishermen of Hastings* (News Books, 1985) tells the complete story.

Hauling crawfish off St Ives aboard Tommy Berriman's *Diligent*, SS6, in 1970. Oilskins had improved by this time whereas oiled canvas had been worn previously. Crawfish nets were large-meshed nets in which they became entangled and had been used after 1945 but their use died out in favour of pots until they were reintroduced in the 1960s. The fishery was confined to a small area between the Lizard and Padstow. (*John McWilliams*)

Bunny Legge of Cadgwith holding a crawfish (*Palinurus elephas*) outside his house in the village. (*Nigel Legge*)

A typical quayside scene at Newquay. The drift-net boats have come in and been beached and the nets are being shaken to clear out the fish, which appear to be pilchards. These are then laced in the round barrels and presumably they will be taken up the beach to a cellar. However it does portray a slightly fanciful scene, something for which fishing is often mistaken. (*Newquay Old Cornwall Society*)

A more serious portrayal of fish being unloaded from a pilchard driver at Port Isaac *c.* 1905. The clothing is more realistic and the men appear more hardened to the job at hand. The barrels were the traditional hogshead barrels used for pilchards and had no belly, just straight sides, as they were used to press the fish in some cellars. Others preferred the half-hogshead barrels.

A postcard dated 1912 of unloading baskets of pilchards at Brazen Island, Polruan. The wicker baskets were often used and were of a standard size.

The quayside at Mevagissey. Here we have quarter-cran baskets which were branded with a government marker that guaranteed their correct size. They were used as a measure of fish and to carry nets and also baited lines.

The quayside at East Looe. Inside the covered market is a pile of fish – possibly unwanted dogfish. The capstans were known as 'ironmen', used to haul in nets, and have been removed from boats, perhaps in readiness for another type of fishing. The trolley with a basket on also appears to have a fish. The quayside is awash with fish too and there's a bustle of activity.

A good catch of fish aboard the Newlyn-registered *Excellent*. Rigged as a trawler at the time, the fish was mostly mackerel although the main catch had already been off-loaded. Often the market for mackerel is so low it ends up unsold. In past times many a farmer used rotting fish to fertilise his land.

A fish sale on Newlyn Beach. Small boats often sold their fish in this way – direct to the customer. The picture is from a painting by Stanhope Forbes.

A postcard view of the fishmarket at Hastings, with boxes and barrels of fish being sold off. Judging by the water on the ground it's a wet day, which might explain why so little people are about.

The crew of the Hastings lugger *Surprise*, RX130, have cleared their drift-nets of herrings in about 1910. The men on the ground are loading the herring into the baskets and are probably merchants as these boats did not have so many men in their crew. Bystanders are also watching, perhaps keen to pick up the odd herring!

The beach at Hastings is a fine example of how fishermen work together to safeguard and maintain the traditions of the beach by operating a cooperative to sell their fish, and they all join in when others need help.

Building a gig on the beach at St Ives, in what was probably a photograph by Henry Trevorrow. The on-lookers do not appear to be the workforce and one wonders what the foreman (he appears so) is telling them. The boat is being traditionally built, and the planking is about to begin though the timber planks alongside look like ties across the heads of the frames – a few are missing. Boats were built with only basic hand tools – adzes, planes, saws, hammers and clamps – whereas now there would be an abundance of electric power tools in use. (*John McWilliams*)

Jackman's Shipyard, Brixham. The yard started in 1850 by the outer breakwater, where this photo was taken in 1880. In 1912 the yard moved next door to Upham's but by 1935 the yard was derelict. Between 1885 and 1926, it is said, Jackmans built some 126 vessels. One of the surviving vessels they built was *Deodar*, BM313. She was sold to Lowestoft in 1919 and thence to Sweden in 1937 and to the present owner in 1972.

Kitto's boatyard in Porthleven, at the head of the harbour. At least two boats are under construction and what appears to be a mast is being formed on the left of the photo. The smoking pots presumably contain molten pitch. If you look carefully there is a number of the workforce looking on. Boatyards were open to anyone to see what was going on: this was long before the days of public liability, when people are no longer accountable for their own stupidity.

The launch of the *Provider*, PZ19, in 1955 from the yard of Oliver & Sons in Porthleven. The yard built seven fishing boats between 1907 and 1961. This boat was the last large fishing boat built at the yard and has more recently been brought back to Porthleven for restoration. (*Carol Williams*)

Two clinker-built boats at Polperro as a fisherman inspects the damage. In 1891, in a gale at Polperro, most of the fleet of a dozen or so luggers were wrecked. The response of many communities these days would be to give up but instead the fishers, having seen such craft in Plymouth (the hookers), turned to gaff-rigged carvel-built vessels, many coming from Looe and Mevagissey boatbuilders. These became known as the Polperro Gaffers. The gaff rig gave better manoeuvrability in confined waters whereas the powerful lug rig gave speed over a long distance. The lug also required more crew while an advantage of it was the lesser amount of rigging.

Vessels needed constant upkeep during the year and here, in Mevagissey, the lugger *Ibis* is having her bottom scrubbed and repainted. The boat is deep at the fore end and there is a propeller aperture in the stern which dates the photo to the 1920s/30s. Boats needed constant maintenance to keep on top of problems and most owners subjected their vessels to an annual overhaul.

Many British fishermen see how the European Union has destroyed their livelihood. Some are not shy to make their point!

Along the Coast

The South Coast is well endowed with small coves, sheltered bays, river estuaries and decent manmade harbours. Cornwall especially is known for its picturesque bays and beaches which were once the haunt of smugglers, with supposed hidden tunnels leading up from the beaches to houses atop the cliff. However different it was in reality, smuggling – or free-trading as many called it – occurred in most parts of the South Coast, where fast boats were able to cross over to France within hours and sail back under cover of darkness. Why, of course, would anyone want to pay the Chancellor a percentage of a bottle of brandy or whatever when it was obtainable at a much reduced price direct from across the water? The same was true for salt – vital to the fish trade – which was heavily taxed in Britain and cheaper in France. Some say the smuggling of salt was what started the whole free trading business. Mind you, the same can be said today when higher taxes are applied this side of the Channel in comparison to those on the other.

Port Isaac has been a busy port since the Middle Ages with stone and slate being exported and coal and timber among the imports. Fishing has also been an important part of the local economy and continues to play its part today. Pilchard fishing was once a huge business and in 1850 there were forty-nine boats and four pilchard cellars, the remains of some of which can still be seen. Here boats lie on the beach before the building of the sea walls.

Kids pose at Padstow. Said to be the only decent port on the north coast of Cornwall, it lies two miles up the River Camel. However, at the mouth of this river is the infamous Doom Bar, which has caught out many a ship of the years – over 600 according to one report. It is also the local fishing registration port, i.e. PW. (*John McWilliams*)

Here dual-purpose steam drifters/trawlers from Lowestoft and Yarmouth lie alongside at Padstow. In the 1920s good hauls of herring were being made in the South West and Padstow, connected to rail network, was a convenient place to land. Trawling in the Irish Sea out of the herring season gave these boats work for much of the year. (*Royal Cornwall Museum, Truro*)

Two trawlers lie alongside at Padstow. By 1908 3,000 tons of fish were being landed and freighted off by train to Billingsgate during the winter fishing. Fish could be taken directly off the boats onto trains that left each afternoon. The First World War killed off much of the trade after German submarines attacked boats. Thirteen were said to have been sunk in one day. (*Jan Pentreath*)

Like Padstow, Newquay's port collapsed after 1918 and tourism slowly followed so that now Newquay is the surf and party centre of the South West as well as a popular holiday centre. It's also famed for its gigs, being the last place to have working gigs. The benchmark gig *Treffry* is also housed in Newquay Rowing Club. But, as the 2008 photograph shows, Newquay still has its own fishing fleet.

The beach landing just below Pendeen Head, on Cornwall's north coast, was built for the pilchard fishery. There's a winch house above and curing sheds atop the slipway. Here, in 2008, several boats were still based for seasonal lobstering and crabbing.

The beach at St Ives, showing several motorised luggers as the tide ebbs. To the left are a couple of gigs and a transom-sterned boat is to the right. Most of the luggers appear to be dandy rigged. (*Richard Major*)

Boats drawn up at Sennen Cove in the 1930s. The large double-ender is a seiner, and it is said that Sennen was once the most important pilchard seine fishery in Cornwall, with 12,000 caught at one time. That is tenuous, however, as this catch is nowhere near comparable to the largest one said to have taken place in St Ives in the autumn of 1851, when it was estimated that 17, 908, 800 pilchards were caught. This catch took a week to land and the overall profit was £7,500, which is over £500,000 in today's sums. Seining at Sennen carried on into the twentieth century and 1,200 stone were said to have been landed at one time in March 1977.

Opposite: Sennen Cove is a very different place today and you'd be forgiven for not realising it was a major fishery station. However, lobsters and crabs have long been an important source of income for Sennen fishers and today several boats still work from there.

Penberth is another tiny hamlet and cove which once had some fifteen working boats, mostly engaged in shellfishing. Today there are still several boats pulled up and four full-time fishers. (*Jan Pentreath*)

Penberth in 2010 with boats drawn up. On the left is the original capstan used to haul the boats up, though today they use an electric winch. The lobster pots are the typical ones of today.

Colourful boats at Penberth in 2010.

A selection of boats at the 2012 Sea, Salts and Sail Festival at Mousehole. Every two years the village holds this festival to celebrate its maritime past. Boats come from all over Cornwall and beyond, and usually a few from Brittany, where similar traditions are not forgotten.

The inner harbour at Newlyn, dating back to the fourteenth century. The harbour grew rapidly in the eighteenth century, with its fleet of luggers and seiners becoming the largest in Cornwall. When Lowestoft drifters tried to land mackerel in May 1896, the boats were attacked and the catch thrown overboard. The drifters left but were chased by several local boats and some were boarded and severe damage inflicted. When the Newlyn folk heard the boats were going to land in nearby Penzance, more than 200 men marched there and confronted the police and coastguards. Hand-to-hand fighting ensued until a gunboat arrived, full of troops, to put an end to the Newlyn Riots.

The same view in 2008. A new pier – the Mary Williams Pier – was opened by the Queen in 1980, increasing the facilities.

Part of the fleet moored against the Mary Williams Pier in 2008. The two inside craft are beam trawlers – *Sara Cathryn Stevenson*, PZ753, and *Bryan D Stevenson*, PZ290 – while next is *Jacqueline*, PZ192, then *Golden Harvest*, PZ63, and *Matthew Harvey*, PZ190. The first three belong to W. Stevenson & Sons of Newlyn.

Decommissioning of fishing boats often meant their enforced scrapping and over the years hundred of perfectly seaworthy wooden vessels have been chopped up at the whim of a government that has no concern about maritime heritage. Here, *Tudor Rose III* is being smashed up by a JCB at Newlyn in 1998.

Above: Another view of Newlyn in 1882, with the old harbour and luggers lined up about where the *Tudor Rose III* was being broken up.

Porthleven harbour with luggers. The church here received media attention, with photographs of spray from the early 2014 storms over it. The harbour, built in the early nineteenth century, was never successful as a commercial port because it faces the onslaught of the Atlantic from the south-west. The fishing fleet didn't mind so much and Porthleven became an important fishing station with at least two boatyards. However, during the storms, the baulks of timber across the harbour were smashed and the surge was so strong that boats were sucked out of the harbour while others were smashed inside it.

Views of Mullion Cove

In 1890, before the harbour was built.

The beach-based boats, with shelter from all but the worst of the south-west storms. (*Jan Pentreath*)

The sea breaking over in January 2014, showing just how vulnerable these small harbours are. (*Luke and Joanna Powell*)

A closer look at the pilchard seiner, this one being the Cover's seiner *Moselle*, photographed in 1934. The square-stern is peculiar and would appear to be broader to aid the shooting of the net. (*Nigel Legge*)

Getting ready for sea – the Cadgwith fleet of crabbers in 1934. The *Minerva* can be seen on the right of the phgotograph. (*Nigel Legge*)

A 1960s postcard view of Cadgwith. Not a lot has changed since the previous photograph although the seiners have gone and the beach seems a bit more orderly!

Coverack was once a fishing port protected by a harbour wall built in 1724. There appears to be a trading vessel sitting forlornly atop the beach, open to the elements with no hatch cover and a crooked topmast. Presumably it has become surplus to requirements and been left abandoned like so many craft of the era. There are a couple of crabbers pulled up the beach and an assortment of other craft. (*Jan Pentreath*)

Across Falmouth Bay is another small fishing port, Portscatho, on the Roseland Peninsula. Built upon the pilchard fishery, there's a harbour wall that offers shelter and several working craft lie in the harbour and pilchard nets hang over the boat. There's the top of a capstan for hauling in the extreme foreground. At the beginning of the eighteenth century there were eleven pilchard cellars in the village, usually fished by the elders as the young men tended to crew the Falmouth packets and merchant vessels. (*Jan Pentreath*)

Portloe at the beginning of the twentieth century, with crab boats moored down the beach. (*Jan Pentreath*)

Portloe in 2003. The beach is steep and two modern crabbers are drawn up. The wooden boat in the middle is similar in shape to the older sailing boats though more full bodied because of the added weight of the engine.

Around Dodman Point is another small harbour, Gorran Haven. Again growing up from being a pilchard port, there is documented evidence that fishers worked from here in 1270. 300 years later, Gorran Haven was a much more important fishery station than Mevagissey. In 1902 the Fowey register had over 200 boats: seventy belonged to Mevagissey, fifty-seven at Looe, forty-three at Gorran Haven and one at Portloe. Today Gorran Haven only has a seasonal crab fishery.

Polkerris, tucked at the top of St Austell Bay was once a busy pilchard harbour with several seines and, in the twentieth century home to a fleet of crabbers. Today it is busy with tourists who come for its fine beach and popular pub. (*Jan Pentreath*)

Polperro harbour, showing the baulks of timber in the foreground and the crane to lift them. These are used to close the harbour in winter when poor weather is forecast. Several harbours along this coast have the same facility – Mousehole being a notable example – and often the harbours are closed off for extended periods over the winter months in adverse weather. Any vessel wanting to go to sea has to use another nearby harbour.

Looe has been a fishing port since medieval times. Here it is the subject of a Victory plywood jigsaw, bought recently in a charity shop (only one piece missing!). Today Looe has some forty-seven fishing vessels, mainly under 10 metres, and 120 fishermen have employment. There is a modern fishmarket and another 40 workers are employed in ancillary related industries.

Next along the coast is Looe, seen here in a photograph taken from the bridge across the river. The fleet of luggers is moored on the east side as usual. During the recent bad weather, Looe has flooded because it does not have any facility to close off the harbour entrance because it is both too wide and the mouth of a river. But the high tides and storm winds meant that the water level went over the quayside.

The last photograph in Cornwall is of the small harbour of Portwrinkle. In 1813 there were simply fishermen's huts along the shore and a curing house, though the whole population of the hamlet would be involved in the pilchard fishery, especially the tucking and curing. A decade earlier some 200 hogsheads – a hogshead was a barrel containing about 3,000 fish – were caught. But in 1834 an advertisement announced the sale by public auction of the equipment and cellars of 'The Portwrickle [sic] Fishery Company', consisting of three seines, boats, tackle and all shore equipment, and 90 tons of new and 30 tons of old salt. Portwrinkle harbour's days were over. (Jan Pentreath)

Sutton Pool was the base for the Plymouth fishing industry. Here a host of Lowestoft drifters are unloading mackerel. Plymouth fishmarket, squashed into the narrow streets of the Barbican, was said to be the best fishmarket in the country. Trawlers also came in to land, and Plymouth had its own fleet of at least thirty sailing trawlers. Today Plymouth has the highest landings of fish in the South West, though this is mostly mackerel.

Crab boats moored off Hope Cove, on the south-east tip of Bigbury Bay. Again it was pilchards and mackerel that led to a harbour being built but Hope became renowned for its crabs latterly. The place has the dubious honour of being the only place in mainland Britain where Spanish sailors landed during the Armada of 1588 after their ship was wrecked. The 140 survivors were at first sentenced to death but later ransomed and returned to Spain. Hope is said to have been the centre of smuggling operations in this area.

Around the other side of the South Hams peninsula is the village of Hallsands at the southern edge of Start Bay. Famed for its shellfish, just as the neighbouring villages of Slapton, Torcross and Beesands were, the beach at Hallsands lies just north of the village itself. Here boats are drawn up on a pleasant summer day. However, Hallsands gained fame when most of the village houses fell into the sea after a gale in 1917 as their foundations had been undermined by the sea, thanks to gravel dredging offshore.

Boats dried out in the harbour at Brixham in 1868. Although there is now a deepwater harbour here, the inner harbour still remains tidal.

The beach at Beer with a hive of activity. Boats are pulled up and, in at least one case, the fish are being shaken from the nets. Barrels are being prepared to be filled and a horse and cart waits to carry the catch away up the hill. Meanwhile, a gaggle of onlookers are enjoying the spectacle.

A boat seen at Portland Bill in 2001. Fishing boats were craned into the sea using the hoist as the whole area is one of cliffs, with the sea 50 feet or so below. The boat is a full-bilged motor boat, with some similarities to the lerret.

The Poole fishing fleet at anchor off the harbour in about 1910. Some have raised sail so presumably they are about to set sail for Poole Bay, where they trawl. It does show there was a considerable fleet here.

Small inshore boats at Emsworth. One boat has what appears to be a beam trawl aboard while the fellow is attending to a net and another boat has a net drying in the rigging. The photographer has obviously attracted attention, for the people on the shore are standing watching.

Bosham from a painting by Stuart Lloyd. The scene gives a feeling of peacefulness – man, woman and child attend to *Daisy of Bosham*, while another man/wife fish off the quay and the crew of two other boats are chatting. It's a far ride from today's Bosham, where houses surround the Norman church and I can find no sign of the thatched cottage any longer.

Three boats on Selsey beach. The fishermen appear to be hauling the net out of the furthermost boat.

The *Rosemary Ann* boarding holidaymakers before they depart for a trip around the bay. The exact place this was taken is not known but the beach is presumed to be Brighton or nearby.

Brighton beach with the Palace Pier in the background, 1958. Brighton beach used to be a thriving fishing location, though tourism created more opportunity for the fishermen in the summer. Two of the boats here have lute sterns, as was the Sussex tradition.

The hive of activity outside the King's Road Arches on Brighton beach *c.* 1890, with luggers and smaller punts sharing the beach with bathing machines. Nets are drying on the walls and sails are drying in the breeze. Today some of the Arches house the Brighton Fisheries Museum.

The punt *Three Brothers* on the beach at Hastings. There's a fisherman's shed just visible on the left and it would seem that the photo was taken to the west of the Stade.

The bulldozer which is used to haul boats up and down the beach.

Various elliptical-sterned boats on the beach at Hastings with seagulls searching for any cast off fish!

The Stade, with its fishermen's huts and net sheds. The Hastings Fishermen's Museum is housed here.

Fish shack at Hastings *c.* 2000, with locally caught fish on sale.

Above: Close up of a hut with a diesel powered winch used to haul boats up the beach.

The net sheds have recently been restored with Heritage Lottery money. Thankfully the Council failed in demolishing these fine examples in the 1940s.

Rye, a couple of miles upstream of the river, is the registration centre here of RX – standing for Rye, Port of Sussex. The fishing fleet was based around the mouth of the river.

FISHING FLEET, FOLKESTONE. 2373.

A postcard of the fishing fleet at Folkestone, showing what a substantial fleet there once was there. Again most of the boats are of the Cornish type and the photo would seem to have been taken early 1940s.

Folkestone from a different aspect as the harbour is full of two-masted luggers like the one entering. This boat, registered as 73FE, has a beam trawl hanging over the starboard side.

The harbour at Ramsgate busy with trawlers.